INTERNATIONAL CENTRE FOR MECHANICAL SCIENCES

COURSES AND LECTURES - No. 76

INGO MUELLER
THE JOHNS HOPKINS UNIVERSITY, BALTIMORE

ENTROPY, ABSOLUTE TEMPERATURE
AND COLDNESS IN THERMODYNAMICS.

BOUNDARY CONDITIONS IN POROUS MATERIALS

COURSE HELD AT THE DEPARTMENT
OF MECHANICS OF SOLIDS
JULY 1971

UDINE 1971

SPRINGER-VERLAG WIEN GMBH

ISBN 978-3-211-81126-9   ISBN 978-3-7091-2965-4 (eBook)
DOI 10.1007/978-3-7091-2965-4

This work is subject to copyright.

All rights are reserved,

whether the whole or part of the material is concerned

specifically those of translation, reprinting, re-use of illustrations,

broadcasting, reproduction by photocopying machine

or similar means, and storage in data banks.

© 1972 by Springer-Verlag Wien

Originally published by Springer Vienna in 1972

PREFACE

Entropy, Absolute Temperature and Coldness in Thermodynamics.

Thermodynamic processes are defined as the solution of the field equations that are based on the equations of balance of mass, momentum and energy and on constitutive relations for stress, heat flux and internal energy. Entropy is introduced as an additive constitutive quantity whose flux is also constitutive and whose production is non-negative. The requirement that this inequality hold for every thermodynamic process leads to some restrictions on the constitutive equations for stress, heat flux and internal energy. Most of these restrictions involve a function of the temperature and its time derivative which, under a suitable continuity assumption for the entropy flux, can be shown to be a universal function. This function will be called the coldness, its equilibrium value is the reciprocal of the absolute temperature.

This new systematic approach to thermodynamics will be illustrated in the first part of this paper for a simple heat conducting fluid and it will be shown, that the theory allows for a finite speed of propagation of disturbances of temperature.

In the second part of the paper the same ideas are applied to rigid heat conducting solids and the speed of propagation of singular surfaces is calculated. This speed turns out to depend on the temperature gradient in general. The paper concludes with the proof that, for a large class of rigid heat conductors, the tensor of heat conductivity is symmetric.

Boundary Conditions in Porous Materials.

The fields in a continuum theory of a fluid in a porous solid are explained to be the densities and the motions of the fluid and the solid and the porosity.

The field equations may be based on the equations of balance of masses, momenta and energy. Darcy's law will be mentioned briefly and jump conditions at interface boundaries will be expounded for various special cases, that are important to hydrologists.

May, 1972

Ingo Müller

Part I

ENTROPY, COLDNESS AND ABSOLUTE TEMPERATURE

1. Thermodynamic Processes.

The main objective of thermodynamics is the determination of the fields of

$$\text{density } \rho(x_k, t),$$
$$\text{velocity } v_i(x_k, t),$$
$$\text{empirical temperature } \vartheta(x_k, t).$$

For this purpose it is customary to rely on the five equations of balance for mass, momentum and energy

$$\frac{\partial \rho}{\partial t} + (\rho v_i)_{,i} = 0,$$

$$\frac{\partial \rho v_i}{\partial t} + (\rho v_j v_i - t_{ji})_{,i} = 0,$$

$$\frac{\partial \rho \varepsilon}{\partial t} + (\rho \varepsilon v_i + q_i)_{,i} = t_{ij} v_{i,j}.$$

These equations must be supplemented by constitutive equations which relate the stress t_{ij}, the heat flux and the specific internal energy ε to the fields ρ, v_i, ϑ in a materially dependent way. In general t_{ij}, q_i, ε could depend on the history of these fields in the whole body, but we consider a very simple case where the constitutive equations have the form

$$t_{ij}(x_k,t) = t_{ij}\left(\rho(x_k,t), v_l(x_k,t), \vartheta(x_k,t), \frac{\partial \vartheta(x_k,t)}{\partial t}, \vartheta_{,l}(x_k,t)\right),$$

$$q_i(x_k,t) = q_i\left(\rho(x_k,t), v_l(x_k,t), \vartheta(x_k,t), \frac{\partial \vartheta(x_k,t)}{\partial t}, \vartheta_{,l}(x_k,t)\right),$$

$$\varepsilon(x_k,t) = \varepsilon\left(\rho(x_k,t), v_l(x_k,t), \vartheta(x_k,t), \frac{\partial \vartheta(x_k,t)}{\partial t}, \vartheta_{,l}(x_k,t)\right).$$

The material that is characterized by these constitutive equations is called a <u>simple heat conducting fluid</u>.

Insertion of the constitutive equations into the equations of balance leads to a determinate system of differential equations from which we may hope to obtain the fields ρ, v_i, ϑ as solutions of well posed initial and boundary value problems. Every solution of this system of field equations is called <u>thermodynamic process</u> in a simple heat conducting fluid.

It ought to be the aim of thermodynamics to find such solutions, however that presupposes that we know the field equations explicitly, that is to say that we know the constitutive equations for a given material. In reality we do not know that and therefore most of thermodynamics is concerned with finding restrictions in the constitutive relations.

2. Restrictive Principles

An important restrictive principle for the constitutive functions postulates frame indifference of these functions. Most people require indifference under general transformations of frame, others are content to assume Galilei invariance of the constitutive functions. In general that makes a difference but not in our particularly simple case. Here both assumptions lead to the result that $v_i(x_k, t)$ and $\frac{\partial \vartheta(x_k,t)}{\partial t}$ can only occur in the combination

$$\dot{\vartheta} \equiv \frac{\partial \vartheta}{\partial t} + v_i \vartheta_{,i}$$

among the variables in the constitutive equations.

Although it would be quite interesting to go deeper into this, I shall not do that here, rather I shall go on to talk about the entropy principle which is the proper subject of this lecture. I assume the following entropy principle:

There exists an additive scalar constitutive quantity, the entropy, with a constitutive flux and with a non-negative production, so that the inequality

$$\frac{\partial \rho \eta}{\partial t} + (\rho \eta v_i + \Phi_i)_{,i} \geq 0$$

holds, η is the specific entropy and Φ_i the entropy flux. The entropy inequality must be satisfied for every thermodynamic process.

In particular, for a simple heat conducting fluid we have

$$\eta = \eta(\rho, \vartheta, \dot{\vartheta}, \vartheta_{,i}),$$

$$\Phi_i = \Phi_i(\rho, \vartheta, \dot{\vartheta}, \vartheta_{,i}).$$

3. Restrictions from the Entropy Inequality on the Constitutive Functions t_{ij}, q_i and ϵ.

Note that the above entropy principle differs from customary ones in that Φ_i is not assumed to be equal to $\frac{q_i}{T}$, where T is the absolute temperature of thermostatics. In fact the absolute temperature has not entered into my arguments so far. While in other thermodynamic theories q_i can be eliminated from the entropy inequality and the balance of internal energy, this shortcut in the search for restrictions on the functions t_{ij}, q_i and ϵ obviously is not open to us, and I shall now proceed to explain how we can obtain such restrictions from the general entropy principle which I propose.

After introducing the constitutive functions for η and Φ_i; in the inequality and carrying out the indicated differentiations we obtain the following explicit form of the entropy inequality

$$\left(\eta + \varrho\frac{\partial\eta}{\partial\varrho}\right)\frac{\partial\varrho}{\partial t} + \varrho\frac{\partial\eta}{\partial\vartheta_{,k}}v_{,k}\frac{\partial v_k}{\partial t} + \varrho\frac{\partial\eta}{\partial\vartheta}\frac{\partial^2\vartheta}{\partial t^2} +$$

$$+ \left[\left(\eta + \varrho\frac{\partial\eta}{\partial\varrho}\right)v_i + \frac{\partial\Phi_i}{\partial\varrho}\right]\varrho_{,i} + \varrho\frac{\partial\eta}{\partial\vartheta}\vartheta + \frac{\partial\Phi_i}{\partial\vartheta}\vartheta_{,i} +$$

$$+ \left[\varrho\eta\delta_{ik} + \left(\varrho\frac{\partial\eta}{\partial\vartheta}v_i + \frac{\partial\Phi_i}{\partial\vartheta}\right)\vartheta_{,k}\right]v_{k,i} +$$

$$+ \left[\varrho\frac{\partial\eta}{\partial\vartheta_{,i}} + \varrho\frac{\partial\eta}{\partial\vartheta}v_i + \frac{\partial\Phi_i}{\partial\vartheta}\right]\frac{\partial\vartheta_{,i}}{\partial t} +$$

$$+ \left[\varrho\left(\frac{\partial\eta}{\partial\vartheta}v_k + \frac{\partial\eta}{\partial\vartheta_{,k}}\right)v_i + \frac{\partial\Phi_i}{\partial\vartheta}v_k + \frac{\partial\Phi_i}{\partial\vartheta_{,k}}\right]\vartheta_{,ik} \geq 0.$$

This inequality, according to the entropy principle, must hold for all processes and in particular therefore for the solution of an initial value problem of the field equations. This statement means for a simple heat conducting fluid that the inaquality has to hold for arbitrary choices of the fields

$$\varrho, \quad v_i, \quad \vartheta, \quad \frac{\partial\vartheta}{\partial t}$$

at the initial time. Now, if these fields are chosen to be analytic, the existence of a unique analytic solution of this initial value problem is proven by the Cauchy-Kowalewsky theorem in the theory of partial differential equations. Therefore there exist many solutions for arbitrary choices of initial values for

$$\rho, \rho_{,i}, v_i, v_{i,k}, \vartheta, \frac{\partial \vartheta}{\partial t}, \vartheta_{,i}, \frac{\partial \vartheta_{,i}}{\partial t}, \vartheta_{,ik}$$

at one point. The time derivatives $\frac{\partial \rho}{\partial t}$, $\frac{\partial v_k}{\partial t}$ and $\frac{\partial^2 \vartheta}{\partial t^2}$ at the initial time however, are related to these arbitrary quantities by the field equations which can be written in the form

$$\frac{\partial \rho}{\partial t} = -v_i \rho_{,i} - \rho \delta_{ik} v_{k,i},$$

$$\frac{\partial v_k}{\partial t} = -v_i v_{k,i} + \frac{1}{\rho} t_{ki,i},$$

$$\frac{\partial^2 \vartheta}{\partial t^2} = \left(v_i v_{k,i} - \frac{1}{\rho} t_{ki,i}\right) \vartheta_{,k} + \frac{1}{(\partial \varepsilon / \partial \vartheta)} \left(-v_k \varepsilon_{,k} + \frac{1}{\rho} t_{ki} v_{k,i} + \right.$$

$$\left. - \frac{1}{\rho} q_{k,k} + \frac{\partial \varepsilon}{\partial \rho} (\rho v_k)_{,k} - \frac{\partial \varepsilon}{\partial \vartheta} \frac{\partial \vartheta}{\partial t} - \frac{\partial \varepsilon}{\partial \vartheta_{,i}} \frac{\partial \vartheta_{,i}}{\partial t}\right).$$

Using these relations we may eliminate $\frac{\partial \rho}{\partial t}$, $\frac{\partial v_k}{\partial t}$ and $\frac{\partial^2 \vartheta}{\partial t^2}$ from the entropy inequality and if the constitutive relations for ε, q_i and t_{ij} are introduced, what results is an inequality that is explicitly linear in the quantities

$$\rho_{,i}, v_{k,i}, \frac{\partial \vartheta_{,i}}{\partial t}, \vartheta_{,ik}.$$

Since, as I have explained, there are trermodynamic processes corresponding to arbitrary choices of these derivatives, we can only satisfy the entropy principle, if none of the terms with $\rho_{,i}$, $v_{k,i}$, $\frac{\partial \vartheta_{,i}}{\partial t}$, and $\vartheta_{,ik}$ contributes to the inequality and from this we infer the following conditions

$$\frac{\partial \Phi_i}{\partial \rho} - \Lambda \frac{\partial q_i}{\partial \rho} = 0 ,$$

$$\frac{\partial \Phi_{(i}}{\partial \vartheta_{,k)}} - \Lambda \frac{\partial q_{(i}}{\partial \vartheta_{,k)}} = 0 , \quad^{(+)}$$

$$\frac{\partial \Phi_i}{\partial \vartheta} - \Lambda \frac{\partial q_i}{\partial \vartheta} + \rho \left(\frac{\partial \eta}{\partial \vartheta_{,i}} - \Lambda \frac{\partial \varepsilon}{\partial \vartheta_{,i}} \right) = 0 ,$$

$$- \rho^2 \left(\frac{\partial \eta}{\partial \rho} - \Lambda \frac{\partial \varepsilon}{\partial \rho} \right) \delta_{ij} + \Lambda t_{ij} + \left(\frac{\partial \Phi_i}{\partial \vartheta} - \Lambda \frac{\partial q_i}{\partial \vartheta} \right) \vartheta_{,i} = 0$$

where the quantity Λ has been introduced according to the definition

$$\frac{\partial \eta}{\partial \vartheta} + \Lambda \frac{\partial \varepsilon}{\partial \vartheta} \equiv 0 .$$

(+) Indices in round brackets indicate that a tensor is symmetrized with respect to these indices.

There remains a residual inequality of the form

$$\rho\left(\frac{\partial \eta}{\partial \vartheta} - \Lambda\frac{\partial \varepsilon}{\partial \vartheta}\right)\dot{\vartheta} + \left(\frac{\partial \Phi_k}{\partial \vartheta} - \Lambda\frac{\partial q_k}{\partial \vartheta}\right)\vartheta_{,k} \geq 0$$

These are restrictions on the constitutive functions for t_{ij}, q_i, ε, η and Φ_i, but we want restrictions on t_{ij}, q_i, and ε only (after all, neither η nor Φ_i occur in our field equations) and some such restrictions can be obtained easily for this simple material, where the constitutive relations must have the form

$$\varepsilon = \varepsilon(\rho, \vartheta, \dot{\vartheta}, g) \quad \text{where} \quad g \equiv \vartheta_{,i}\vartheta_{,i}$$

$$\eta = \eta(\rho, \vartheta, \dot{\vartheta}, g)$$

$$q_i = -K\vartheta_{,i}$$

$$\Phi_i = \varphi\vartheta_{,i}$$

$$t_{ij} = -p\delta_{ij} + Q\vartheta_{,i}\vartheta_{,j}, \text{ where } K, \varphi, p, Q \text{ may all depend on } \rho, \vartheta, \dot{\vartheta}, g$$

K is called the heat conductivity and p is called the pressure. With $\Phi_i = \varphi\vartheta_{,i}$, and $q_i = -K\vartheta_{,i}$ the relation $\frac{\partial \Phi_{(i}}{\partial \vartheta_{,j)}} - \Lambda\frac{\partial q_{(i}}{\partial \vartheta_{,j)}} = 0$ assumes the form

$$(\varphi + \Lambda K)\delta_{ij} + 2\left(\frac{\partial \varphi}{\partial g} + \Lambda\frac{\partial K}{\partial g}\right)\vartheta_{,i}\vartheta_{,j} = 0$$

whence we conclude that

$$\varphi = -\Lambda K, \qquad \frac{\partial \Lambda}{\partial g} = 0 \text{ so that } \Phi_i = \Lambda(\varrho, \vartheta, \dot{\vartheta}) q_i.$$

In fact Λ is also independent of ϱ as we can easily infer from our relation $\frac{\partial \Phi_i}{\partial \varrho} - \Lambda \frac{\partial q_i}{\partial \varrho} = 0$. Thus we have

$$\Phi_i = \Lambda(\vartheta, \dot{\vartheta}) q_i$$

and if we use this, the remainder of the restrictive conditions can be written as

$$\frac{\partial \eta}{\partial \dot{\vartheta}} = \Lambda \frac{\partial \varepsilon}{\partial \dot{\vartheta}},$$

$$\frac{\partial \eta}{\partial g} = \Lambda \frac{\partial \varepsilon}{\partial g} + \frac{K}{2\varrho} \frac{\partial \Lambda}{\partial \vartheta} = \Lambda \left(\frac{\partial \varepsilon}{\partial g} + \frac{Q}{2\varrho} \right),$$

$$\frac{\partial \eta}{\partial \varrho} = \Lambda \frac{\partial \varepsilon}{\partial \varrho} - \Lambda \frac{\eta}{\varrho^2}$$

$$\left(\frac{\partial \eta}{\partial \vartheta} - \Lambda \frac{\partial \varepsilon}{\partial \vartheta} \right) \dot{\vartheta} + \frac{\partial \Lambda}{\partial \vartheta} q_i \vartheta_{,i} \geq 0.$$

From the first three of these relations we obtain integrability conditions which do not contain η any longer:

(I) $\qquad \dfrac{\partial \ln \Lambda}{\partial \vartheta} = \dfrac{2 \dfrac{\partial p}{\partial g}}{K + \varrho \dfrac{\partial K}{\partial \varrho}} = -\dfrac{\dfrac{\partial p}{\partial \vartheta}}{p - \varrho^2 \dfrac{\partial \varepsilon}{\partial \varrho}} = \dfrac{Q}{K},$

(II)
$$\frac{\partial}{\partial \dot{\vartheta}}\left(\ln \frac{\partial \Lambda}{\partial \dot{\vartheta}}\right) = -\frac{\frac{\partial K}{\partial \dot{\vartheta}} + 2\rho \frac{\partial \varepsilon}{\partial \dot{g}}}{K}$$

We conclude that the expressions in the right hand sides of (I) and (II) are functions of ϑ and $\dot{\vartheta}$ only.

More can be learned from the residual inequality. Its left hand side is a function $\sigma(\rho, \vartheta, \dot{\vartheta}, \vartheta_{,i})$ which obviously has its minimum value, namely zero, in a time independent and uniform process which I call <u>equilibrium</u> so that

$$\sigma|_E = \sigma(\rho, \vartheta, 0, 0) = 0.$$

Of necessity therefore we must have

$$\frac{\partial \sigma}{\partial \dot{\vartheta}}\bigg|_E = 0, \quad \frac{\partial \sigma}{\partial \vartheta_{,i}}\bigg|_E = 0 \quad \text{and} \quad \begin{vmatrix} \frac{\partial^2 \sigma}{\partial \dot{\vartheta}^2}\big|_E & \frac{\partial^2 \sigma}{\partial \vartheta_{,i}\partial \dot{\vartheta}}\big|_E \\ \frac{\partial^2 \sigma}{\partial \dot{\vartheta}\partial \vartheta_{,k}}\big|_E & \frac{\partial^2 \sigma}{\partial \vartheta_{,i}\partial \vartheta_{,k}}\big|_E \end{vmatrix} \text{non-negative definite.}$$

From the first one of these requirements we obtain rather obviously $\frac{\partial \eta|_E}{\partial \vartheta} = \Lambda|_E \frac{\partial \varepsilon|_E}{\partial \vartheta}$ which combines with a previous relation for $\frac{\partial \eta}{\partial \rho}$ to give

$$d\eta|_E = \Lambda|_E \left(\frac{\partial \varepsilon|_E}{\partial \vartheta} d\vartheta + \left(\frac{\partial \varepsilon|_E}{\partial \rho} + \frac{p|_E}{\rho^2}\right) d\rho\right),$$

(III)
$$d\eta|_E = \Lambda|_E \left(d\varepsilon|_E + \frac{p|_E}{\rho^2} d\rho \right).$$

I withhold an obvious comment on the similarity of this equation with Gibbs' equation of thermostatics and list the implications of the requirement that the above matrix be non-negative definite. These implications are the inequalities

(IV) $\quad \dfrac{\partial \Lambda|_E}{\partial \vartheta} K|_E \leq 0 \quad$ and $\quad \dfrac{\partial \Lambda|_E}{\partial \vartheta} \dfrac{\partial \varepsilon}{\partial \vartheta}\bigg|_E \geq \dfrac{\partial \Lambda}{\partial \vartheta}\bigg|_E \dfrac{\partial \varepsilon|_E}{\partial \vartheta}$

which I shall interpret later.

The relations (I) through (IV) summarize whatever restrictive conditions we can obtain from the entropy principle alone. It is only (I) that expresses some restrictions of the required kind, that is, restrictions on the form of the functions t_{ij}, q_i and ε. The relations (II') through (IV) all contain Λ which is defined in terms of η. Let us learn more about Λ:

4. The Coldness.

We describe a thin wall between two different simple heat conducting fluids I and II as a material singular surface. Under rather general assumptions the balance of internal energy at such a surface reduces to the statement that the normal component of the heat flux be continuous across the surface.

$[\vartheta(t)] \stackrel{\wedge}{=}$ ideal wall $[q_\perp] = 0$ or $q_\perp^I = q_\perp^{II}$.

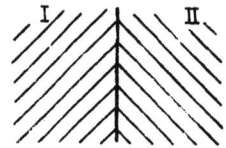
We call the wall <u>ideal</u>, if for all processes the jump of the temperature across the wall vanishes, and now we <u>assume</u> that at an ideal wall the normal component of the entropy flux is also continuous so that

$$[\Phi_\perp] = 0 \quad \text{if} \quad [\vartheta(t)] = 0 .$$

By use of $\Phi_i = \Lambda(\vartheta, \dot{\vartheta})q_i$ and of $[q_\perp] = 0$ we infer from this assumption that

$$\Lambda^I(\vartheta, \dot{\vartheta}) = \Lambda^{II}(\vartheta, \dot{\vartheta})$$

holds, i.e. <u>Λ is a universal function of the temperature and its time derivative, it is called coldness.</u>

Of course, $\Lambda|_E$ is then a universal function of the temperature alone and from (III) we conclude that <u>$\Lambda|_E = \Lambda(\vartheta, 0)$ is the reciprocal of the absolute temperature of thermostatics:</u> <u>$\Lambda|_E = 1/T(\vartheta)$</u> . The proof of this statement is simple, because $\frac{1}{T(\vartheta)}$ is defined in thermostatics as <u>the</u> integrating factor of $d\varepsilon|_E + \frac{p|_E}{\varrho^2}d\varrho$ that is universal and a function of ϑ alone. From (I) and (II) we may now conclude that the expressions

$$\frac{2\frac{\partial p}{\partial g}}{K - \rho \frac{\partial K}{\partial \rho}} \quad , \quad -\frac{\frac{\partial p}{\partial \dot{\vartheta}}}{p + \rho^2 \frac{\partial \varepsilon}{\partial \rho}} \quad , \quad \frac{Q}{K}$$

are universal functions of ϑ and $\dot{\vartheta}$, apart from being equal, which we knew before, and that

$$\frac{\frac{\partial K}{\partial \dot{\vartheta}} + 2\rho \frac{\partial \varepsilon}{\partial g}}{K}$$

is also a universal function of ϑ and $\dot{\vartheta}$. The coldness may be calculated as a function of ϑ and $\dot{\vartheta}$ from either one of the equations (I) or (II) after measurements of p and K (say) have been made for just <u>one</u> simple heat conducting fluid. With $\Lambda|_E = 1/T(\vartheta)$ we may rewrite the relation (IV) in the form

$$\frac{1}{T^2}\frac{dT}{d\vartheta} K|_E \geq 0 \quad \text{and} \quad -\frac{1}{T^2}\frac{dT}{d\vartheta}\frac{\partial \varepsilon}{\partial \dot{\vartheta}}\bigg|_E \geq \frac{\partial \varepsilon|_E}{\partial \vartheta}\frac{\partial \Lambda}{\partial \vartheta}\bigg|_E .$$

We know from thermostatics that the absolute temperature is a non-vanishing monotonically increasing function of ϑ and therefore we obtain

$$K|_E \geq 0 \quad \text{and} \quad \frac{\partial \varepsilon}{\partial \dot{\vartheta}}\bigg|_E \leq \frac{T^2}{\frac{dT}{d\vartheta}} \frac{\partial \varepsilon|_E}{\partial \vartheta} \frac{\partial \Lambda}{\partial \vartheta}\bigg|_E .$$

The first inequality is a standard result in irreversible thermodynamics, but the second one is not and I shall presently discuss one of its implications.

Of course, statements like the one about the universal character of $\frac{\partial p/\partial g}{K-\rho\partial K/\partial\rho}$ etc. may be the subject to experimental verification in principle. However, most of these results concern the coefficients of terms in the constitutive equations for t_{ij}, q_i and ε that are non-linear in ϑ and $\dot{\vartheta}$ and, so far to my knowledge, such terms have not been accessible to thermodynamic experimentation. Therefore, at present, we must conclude that the concept of coldness is wanting corroboration from experiments.

5. The Speed of Heat Propagation.

Let us consider a simple heat conducting fluid at rest and with uniform density. The only relevant field in that case is the field of temperature and the only relevant field equation is the balance of internal energy which has the form

$$\rho\frac{\partial\varepsilon}{\partial t} + q_{i,i} = 0 \;.$$

The constitutive equation read

$$\varepsilon = \varepsilon\left(\vartheta, \frac{\partial\vartheta}{\partial t}, g\right)$$

and

$$q_i = -K\left(\vartheta, \frac{\partial\vartheta}{\partial t}, g\right)\vartheta_{,i} \;.$$

The speed of heat propagation

We introduce these into the above balance and linearize around equilibrium thus obtaining a linear equation of heat conduction of the form

$$\rho \frac{\partial \varepsilon}{\partial \vartheta}\Big|_E \frac{\partial \vartheta}{\partial t} + \rho \frac{\partial \varepsilon}{\partial \dot{\vartheta}}\Big|_E \frac{\partial^2 \vartheta}{\partial t^2} = K|_E \vartheta_{,ii} \;.$$

The underlined part of this equation is the usual parabolic equation of heat conduction, according to which a thermal disturbance at any point in the body is felt instantly at every other point. The additional term in the above equation may make the equation hyperbolic so that thermal disturbances propagate with the finite speed

$$u = \sqrt{\frac{K|_E}{\rho \, \partial \varepsilon/\partial \dot{\vartheta}|_E}} \;.$$

Of course, the equation is only hyperbolic, if

$$\frac{\partial \varepsilon}{\partial \dot{\vartheta}}\Big|_E > 0 \;;$$

the point is that this inequality may indeed hold in this theory because we have the relation

$$\frac{\partial \varepsilon}{\partial \dot{\vartheta}} \leq \frac{T^2}{dT/d\vartheta} \frac{\partial \varepsilon}{\partial \vartheta}\Big|_E \frac{\partial \Lambda}{\partial \dot{\vartheta}}\Big|_E$$

so that $\frac{\partial \varepsilon}{\partial \dot{\vartheta}}\Big|_E$ may indeed be positive. This is a genuine improvement on the customary thermodynamic theory which may be obtained by replacing the coldness, wherever it occurs, by its equilibrium value $\frac{1}{T(\vartheta)}$. Were we to accept this, the last inequality would read $\left.\frac{\partial \varepsilon}{\partial \vartheta}\right|_E \leq 0$ and we should obtain a parabolic equation of heat conduction or, even worse, an elliptic one.

Part II

ON HEAT CONDUCTION IN A RIGID BODY

1. General Ideas.

In my paper "The Coldness, a Universal Function in Thermodynamics of Simple Heat Conducting Fluids" I have investigated the consequences of an entropy principle which may roughly be described as follows:

> In every body there exists a scalar additive quantity, the entropy, whose production density is non-negative for every thermodynamic process and whose density and flux are constitutive quantities.

This general entropy principle was supplemented by a continuity condition for the normal component of the entropy flux at ideal walls, i.e. walls which permit the temperature to be continuous.

These ideas led to the notion of the coldness, a universal function of the temperature and the time derivative of the temperature whose equilibrium value is the reciprocal of the absolute temperature of thermostatics. Just like in thermostatics the universal character of absolute temperature leads to universal relations between the pressure and the internal energy, the universal character of the coldness implies universal relations between thermodynamic properties of stress, heat flux and internal energy.

The above ideas are also applicable to more complicated materials and I have been able to prove the existence of the coldness for viscous heat conducting fluids (+) and for thermoelastic solids (++).

When the constitutive relations become more elaborate, the necessary analysis grows more and more complicated and therefore in this paper I restrict the attention again to a very simple material, namely a rigid solid for which I shall be able to predict some interesting results on the theory of heat conduction.

Before I embark on this, however, I should like to mention that from recent research there are indications that neither the concept of coldness nor that of absolute temperature carry over to thermodynamics of such materials as relativistic fluids or mixtures of fluids. I believe that this fact implies some potentially interesting and as yet unexplored properties of matter and energy

---

(+) I.MÜLLER, Die Kältefunktion, eine universelle Funktion in der Thermodynamik visköser warmeleitender Flüssigkeiten. Arch. Rat. Mech. Anal. 40, p.1.
(++) I.MÜLLER, The Coldness, a Universal Function in Thermodynamics of Thermoelastic Solids, Arch.Rat.Mech.Anal. 41 p. 319.

2. Restrcitions from the Entropy Inequality on the Constitutive Relations for Rigid Solids.

The only thermodynamic field of interest in rigid solids is the field of the (empirical) temperature ϑ and the corresponding field equation is commonly based on the balance of internal energy

$$\varrho \frac{\partial \varepsilon}{\partial t} + q_{i,i} = 0$$

where ϱ is the constant density, ε is the specific internal energy and q_i is the heat flux. ε and q_i must be given by constitutive equation which for a rigid solid have the form

$$\varepsilon = \varepsilon\left(\vartheta, \frac{\partial \vartheta}{\partial t}, \vartheta_{,i}\right),$$

$$q_i = q_i\left(\vartheta, \frac{\partial \vartheta}{\partial t}, \vartheta_{,i}\right).$$

The field equation for ϑ results by introducing the constitutive equations into the balance of internal energy and every field $\vartheta(\lambda_i, t)$ that satisfies this equation is called a <u>thermodynamic process in a rigid solid</u>.

The constitutive relations for ε and q_i are restricted by the entropy principle:

There exists a scalar additive quantity, the entropy,

whose production density is non-negative so that the entropy inequality

$$\rho \frac{\partial \eta}{\partial t} + \Phi_{i,i} \geq 0$$

holds for every thermodynamic process. The specific entropy and the entropy flux are given by the constitutive relations

$$\eta = \eta\left(\vartheta, \frac{\partial \vartheta}{\partial t}, \vartheta_{,i}\right),$$

$$\Phi_i = \Phi_i\left(\vartheta, \frac{\partial \vartheta}{\partial t}, \vartheta_{,i}\right).$$

More explicitly the entropy inequality reads

$$\rho \frac{\partial \eta}{\partial \vartheta} \frac{\partial \vartheta}{\partial t} + \rho \frac{\partial \eta}{\partial \dot\vartheta} \frac{\partial^2 \vartheta}{\partial t^2} + \rho \frac{\partial \eta}{\partial \vartheta_{,i}} \frac{\partial \vartheta_{,i}}{\partial t} + \frac{\partial \Phi_i}{\partial \vartheta} \vartheta_{,i} + \frac{\partial \Phi_i}{\partial \dot\vartheta} \frac{\partial \vartheta_{,i}}{\partial t} + \frac{\partial \Phi_i}{\partial \vartheta_{,k}} \vartheta_{,ik} \geq 0 \; ^{(+)}$$

This inequality has to hold for all thermodynamic processes, that is to say, here, for all solutions of the energy equation

$$\rho \frac{\partial \varepsilon}{\partial \vartheta} \frac{\partial \vartheta}{\partial t} + \rho \frac{\partial \varepsilon}{\partial \dot\vartheta} \frac{\partial^2 \vartheta}{\partial t^2} + \rho \frac{\partial \varepsilon}{\partial \vartheta_{,i}} \frac{\partial \vartheta_{,i}}{\partial t} + \frac{\partial q_i}{\partial \vartheta} \vartheta_{,i} + \frac{\partial q_i}{\partial \dot\vartheta} \frac{\partial \vartheta_{,i}}{\partial t} + \frac{\partial q_i}{\partial \vartheta_{,k}} \vartheta_{,ik} = 0.$$

(+) For simplicity in notation I write $\dfrac{\partial \eta}{\partial \dot\vartheta}$ instead of $\dfrac{\partial \eta}{\partial \frac{\partial \vartheta}{\partial t}}$.

Elimination of $\frac{\partial^2 \vartheta}{\partial t^2}$ from the last two relations leads to the inequality

$$\rho\left(\frac{\partial \eta}{\partial \vartheta} - \Lambda \frac{\partial \varepsilon}{\partial \vartheta}\right)\frac{\partial \vartheta}{\partial t} + \left(\frac{\partial \Phi_i}{\partial \vartheta} - \Lambda \frac{\partial q_i}{\partial \vartheta}\right)\vartheta_{,i} - \left(\rho\left(\frac{\partial \eta}{\partial \vartheta_{,i}} - \Lambda \frac{\partial \varepsilon}{\partial \vartheta_{,i}}\right) - \left(\frac{\partial \Phi_i}{\partial \vartheta} - \Lambda \frac{\partial q_i}{\partial \vartheta_{,k}}\right)\right)\frac{\partial \vartheta_{,i}}{\partial t} +$$
$$+ \left(\frac{\partial \Phi_i}{\partial \vartheta_{,k}} - \Lambda \frac{\partial q_i}{\partial \vartheta_{,k}}\right)\vartheta_{,ik} \geq 0,$$

where I have introduced the abbreviation Λ by the definition

$$\frac{\partial \eta}{\partial \vartheta} = \Lambda \frac{\partial \varepsilon}{\partial \vartheta} \quad . \tag{I}$$

Now, since according to the Cauchy Kowalewski theorem of the theory of partial differential equations, there exist solutions of the energy equation for arbitrary choices of

$$\frac{\partial \vartheta}{\partial t} \; , \; \vartheta_{,i} \; , \; \frac{\partial \vartheta_{,i}}{\partial t} \; , \; \vartheta_{,ik}$$

at one point at the initial time, we can only satisfy the entropy principle, if the following conditions hold:

$$\frac{\partial \Phi_{(i}}{\partial \vartheta_{,k)}} - \Lambda \frac{\partial q_{(i}}{\partial \vartheta_{,k)}} = 0 \quad , \quad ^{(+)} \tag{II}$$

$$\rho\left(\frac{\partial \eta}{\partial \vartheta_{,i}} - \Lambda \frac{\partial \varepsilon}{\partial \vartheta_{,i}}\right) + \left(\frac{\partial \Phi_i}{\partial \vartheta} - \Lambda \frac{\partial q_i}{\partial \vartheta}\right) = 0 \quad , \tag{III}$$

$$\rho\left(\frac{\partial \eta}{\partial \vartheta} - \Lambda \frac{\partial \varepsilon}{\partial \vartheta}\right)\frac{\partial \vartheta}{\partial t} + \left(\frac{\partial \Phi_i}{\partial \vartheta} - \Lambda \frac{\partial q_i}{\partial \vartheta}\right)\vartheta_{,i} \geq 0. \tag{IV}$$

(+) Indices in round brackets indicate symmetrization.

3. Isotropic Rigid Solids and the Coldness.

We must realize that the relations (I) through (IV) restrict the constitutive relations for ε, q_i, η and Φ_i whereas we are interested in restrictions on ε and q_i only (after all neither η nor Φ_i occur in our field equation), and such restrictions are easily obtained for <u>isotropic</u> rigid solids where the constitutive relations must have the forms

$$\varepsilon = \varepsilon\left(\vartheta, \frac{\partial \vartheta}{\partial t}, g\right), \text{ where } g = \vartheta_{,i}\vartheta_{,i}$$

$$q_i = -K\left(\vartheta, \frac{\partial \vartheta}{\partial t}, g\right)\vartheta_{,i},$$

$$\eta = \eta\left(\vartheta, \frac{\partial \vartheta}{\partial t}, g\right),$$

$$\Phi_i = \varphi\left(\vartheta, \frac{\partial \vartheta}{\partial t}, g\right)\vartheta_{,i}.$$

K is called the heat conductivity.
Introducing $q_i = -K\vartheta_{,i}$ and $\Phi_i = \varphi\vartheta_{,i}$ into (II) we obtain

$$(\varphi + \Lambda K)\delta_{ik} + \left(\frac{\partial \varphi}{\partial g} + \Lambda\frac{\partial K}{\partial g}\right)\vartheta_{,i}\vartheta_{,k} = 0,$$

whence we conclude that

$$\Phi_i = \Lambda\left(\vartheta, \frac{\partial \vartheta}{\partial t}\right)q_i.$$

If that is inserted in (I), (II) and (IV), we get

$$\frac{\partial \eta}{\partial \vartheta} = \Lambda \frac{\partial \varepsilon}{\partial \vartheta} ,$$

$$\frac{\partial \eta}{\partial g} = \Lambda \frac{\partial \varepsilon}{\partial g} + \frac{K}{2\varrho} \frac{\partial \Lambda}{\partial \vartheta} ,$$

$$\varrho \left(\frac{\partial \eta}{\partial \vartheta} - \Lambda \frac{\partial \varepsilon}{\partial \vartheta} \right) \frac{\partial \vartheta}{\partial t} + \left(\frac{\partial \Phi_i}{\partial \vartheta} - \Lambda \frac{\partial q_i}{\partial \vartheta} \right) \vartheta_{,i} \geq 0 ,$$

and from the first two of these equations it follows by cross differentiation that

$$\frac{\partial}{\partial \vartheta} \left(\ln \frac{\partial \Lambda}{\partial \vartheta} \right) = - \frac{\frac{\partial K}{\partial \vartheta} + 2\varrho \frac{\partial \varepsilon}{\partial g}}{K} . \qquad (V)$$

Furthermore, just like in the case of a simple heat conducting fluid we infer from the inequality that in equilibrium, i.e. in a uniform and time independent process we have

$$\frac{\partial \eta|_E}{\partial \vartheta} = \Lambda|_E \frac{\partial \varepsilon|_E}{\partial \vartheta} , \quad -K|_E \frac{\partial \Lambda|_E}{\partial \vartheta} \geq 0 , \quad \frac{\partial \Lambda|_E}{\partial \vartheta} \frac{\partial \varepsilon}{\partial \vartheta}\bigg|_E - \frac{\partial \Lambda}{\partial \vartheta}\bigg|_E \frac{\partial \varepsilon|_E}{\partial \vartheta} \geq 0 \qquad (VI)$$

While (V) and (VI) represent some restrictions on ε and K, e.g. that the combination $\frac{1}{K}\left(\frac{\partial K}{\partial \vartheta} + 2\varrho \frac{\partial \varepsilon}{\partial g} \right)$ is a function of ϑ and $\frac{\partial \vartheta}{\partial t}$ only, we are able to learn more by what seems to be a safe assumption on Φ_i.

We consider an ideal wall (+) between two different materials, a rigid solid I and a simple heat conducting fluid II. The normal component of the heat flux is continuous under rather general conditions and we assume that at an ideal wall, the normal component of the entropy flux does not jump either. Hence with $\Phi_i = \Lambda\left(\vartheta, \frac{\partial \vartheta}{\partial t}\right) q_i$ we obtain

$$\Lambda^I\left(\vartheta, \frac{\partial \vartheta}{\partial t}\right) = \Lambda^{II}\left(\vartheta, \frac{\partial \vartheta}{\partial t}\right).$$

That is to say: $\underline{\Lambda \text{ is a universal functions of } \vartheta \text{ and } \frac{\partial \vartheta}{\partial t}, \text{ it}}$ $\underline{\text{is called the coldness.}}$

I have already proved for simple heat conducting materials that $\Lambda|_E \geq \Lambda(\vartheta, 0)$ is the reciprocal of the absolute temperature $T(\vartheta)$ of thermostatics, and this result carries over, of course, to rigid solids.

From (V) we can now conclude that $\frac{1}{K}\left(\frac{\partial K}{\partial \vartheta} + 2\varrho \frac{\partial \varepsilon}{\partial g}\right)$ is a universal function of ϑ and $\frac{\partial \vartheta}{\partial t}$ and $(VI)_{2,3}$ can be written as

(VII) $K|_E \geq 0$ and $\frac{\partial \varepsilon}{\partial \vartheta}\Big|_E \leq \frac{T^2}{dT/d\vartheta} \frac{\partial \varepsilon}{\partial \vartheta}\Big|_E \frac{\partial \Lambda}{\partial \vartheta}\Big|_E.$

(+) As in my paper on simple heat conducting fluids I call a wall ideal, if it does not permit a jump in the temperature.

4. Temperature Waves in Isotropic Rigid Solids.

I propose to calculate the velocity of propagation of a wave in the second derivatives of temperature in an isotropic rigid solid. The jumps $[\vartheta]$, $\left[\dfrac{\partial \vartheta}{\partial t}\right]$ and $[\vartheta_{,i}]$ across the wave vanish and therefore the constitutive quantities ε and q_i are also continuous across the wave. Of course, the balance of energy holds on either side so that we obtain

$$\rho\frac{\partial \varepsilon}{\partial \vartheta}\left[\frac{\partial^2 \vartheta}{\partial t^2}\right] + \left(2\rho\frac{\partial \varepsilon}{\partial g} - \frac{\partial K}{\partial \vartheta}\right)\vartheta_{,i}\left[\frac{\partial \vartheta_{,i}}{\partial t}\right] - \left(K\delta_{ij} + 2\frac{\partial K}{\partial g}\vartheta_{,i}\vartheta_{,j}\right)[\vartheta_{,ij}] = 0.$$

If indeed $[\vartheta] = 0$, $\left[\dfrac{\partial \vartheta}{\partial t}\right] = 0$ and $[\vartheta_{,i}] = 0$ we know from the theory of moving singular surfaces that the jumps of the second derivatives of ϑ are all determined by two parameters of which one is the normal speed $(u_e e_e)$ of the surface. We have

$$\left[\frac{\partial^2 \vartheta}{\partial t^2}\right] = C(u_e e_e)^2, \quad \left[\frac{\partial \vartheta_{,i}}{\partial t}\right] = -C(u_e e_e)e_i, \quad \text{where } C = [\vartheta_{,ik}]e_i e_k$$

and e_i is the wave normal. When these <u>compatibility conditions</u> are inserted in the above balance of energy, we are led to a quadratic equation for the velocity $(u_e e_e)$ whose solution has the form

$$(\omega_e e_e) = \frac{2\rho\frac{\partial \varepsilon}{\partial g} - \frac{\partial K}{\partial \vartheta}}{2\rho\frac{\partial \varepsilon}{\partial \vartheta}}(\vartheta_{,i} e_i) \pm \sqrt{\left(\frac{2\rho\frac{\partial \varepsilon}{\partial g} - \frac{\partial K}{\partial \vartheta}}{2\rho\frac{\partial \varepsilon}{\partial \vartheta}}\right)^2 (\vartheta_{,i} e_i)^2 + \frac{K + 2\frac{\partial K}{\partial g}(\vartheta_{,i} e_i)^2}{\rho\frac{\partial \varepsilon}{\partial \vartheta}}}.$$

It follows that the speed of propagation depends on the temperature gradient and we conclude that

$$|\overrightarrow{\omega_e e_e}| - |\overleftarrow{\omega_e e_e}| = \frac{2\rho\frac{\partial \varepsilon}{\partial g} - \frac{\partial K}{\partial \vartheta}}{\rho\frac{\partial \varepsilon}{\partial \vartheta}}|\vartheta_{,i} e_i|,$$

where $|\overrightarrow{\omega_e e_e}|$ is the speed of a wave moving into a region of high temperature, whereas into a region with lower temperature the wave moves with the speed $|\overleftarrow{\omega_e e_e}|$.

Of course, if $\vartheta_{,i} e_i$ is zero, we get

$$|\omega_e e_e| = \sqrt{\frac{K}{\rho\frac{\partial \varepsilon}{\partial \vartheta}}}.$$

This result is a slight generalization of the final result in my paper on a simple heat conduction fluid which was derived within a linear theory of heat conduction and here again we see that this new approach to thermodynamics allows for finite speeds of propagation of temperature waves, since (VII) permits both K and $\frac{\partial \varepsilon}{\partial \vartheta}$ to be positive. The above results hold for waves of second derivatives of ϑ, but formally identical results can be derived for waves of first derivatives of ϑ provided their amplitude is small enough.

5. Symmetry of the Tensor of Heat Conductivity in Anisotropic Rigid Solids.

As the properties of more and more materials are being investigated with the thermodynamic theory presented here, it becomes more and more irritating that I do not know in general the consequences of the differential equations

$$\frac{\partial \Phi_{(i}}{\partial \vartheta_{,k)}} - \Lambda \frac{\partial q_{(i}}{\partial \vartheta_{,k)}} = 0$$

for the forms of Λ and Φ_i. In fact so far only isotropic materials have been dealt with in full generality and when I now consider an anisotropic rigid solid, I shall only get definite results, if I restrict the attention to a special case that is characterized by constitutive relations for q_i and Φ_i which are linear in $\vartheta_{,i}$. Accordingly I assume that the following constitutive relations hold

$$q_i = - K_{ij}\left(\vartheta, \frac{\partial \vartheta}{\partial t}\right) \vartheta_{,j} \quad , \quad \Phi_i = \varphi_{ij}\left(\vartheta, \frac{\partial \vartheta}{\partial t}\right) \vartheta_{,j} \quad , \quad ^{(+)}$$

(+) While these relations obviously represent a special choice, I wish to emphasize that they are still general, if compared with the equations of linear irreversible thermodynamics where people are very much concerned with the symmetry of the tensor of heat conductivity.

whereas ε and η are given, as before, by

$$\varepsilon = \varepsilon\left(\vartheta, \frac{\partial \vartheta}{\partial t}, \vartheta_{,i}\right),$$

$$\eta = \eta\left(\vartheta, \frac{\partial \vartheta}{\partial t}, \vartheta_{,i}\right).$$

K_{ij} is called the tensor of heat conductivity.

The equations (II) imply trivially that Λ is independent of $\vartheta_{,i}$, so that we have $\Lambda = \Lambda\left(\vartheta, \frac{\partial \vartheta}{\partial t}\right)$. Also from (II) it can easily be concluded that

$$\varphi_{(ij)} + \Lambda K_{(ij)} = 0$$

and hence we have

$$\Phi_i = -\Lambda(K_{(ij)} + K_{[ij]})\vartheta_{,j} + (\varphi_{[ij]} + \Lambda K_{[ij]})\vartheta_{,j} , \quad^{(+)}$$

$$\Phi_i = \Lambda q_i + (\varphi_{[ij]} + \Lambda K_{[ij]})\vartheta_{,j} .$$

To reduce this relation further we consider again an ideal wall between an isotropic rigid solid I and an anisotropic one II and assume as before, that

$$[q_i e_i] = 0 \text{ and that } [\Phi_i e_i] = 0, \text{ if } [\vartheta(t)] = 0.$$

For I we have already seen in Section 3 (see page 26) that $\Phi_i = \Lambda\left(\vartheta, \frac{\partial \vartheta}{\partial t}\right) q_i$, while in II we have the above relation.

(+) Indices in square brackets indicate antisymmetrization.

Hence it follows that

$$\Lambda^I\left(\vartheta, \frac{\partial \vartheta}{\partial t}\right) - \Lambda^{II}\left(\vartheta, \frac{\partial \vartheta}{\partial t}\right) = (\varphi_{[ij]} + \Lambda K_{[ij]}) \frac{\vartheta_{,j} e_i}{q_i e_i}.$$

The left hand side of this equation does not depend on $\vartheta_{,i}$ so the right hand side must not depend on $\vartheta_{,i}$ either. It follows therefore that $\varphi_{[ij]} + \Lambda K_{[ij]} = 0$ so that the relation

$$\Phi_i = \Lambda\left(\vartheta, \frac{\partial \vartheta}{\partial t}\right) q_i$$

holds just as in an isotropic rigid solid.

If $\Phi_i = \Lambda q_i$ is introduced into (III), what results is

$$\varrho\left(\frac{\partial \eta}{\partial \vartheta_{,i}} - \Lambda \frac{\partial \varepsilon}{\partial \vartheta_{,i}}\right) = \frac{\partial \Lambda}{\partial \vartheta} q_i$$

and hence we obtain by differentiation with respect to $\vartheta_{,k}$

$$\varrho\left(\frac{\partial^2 \eta}{\partial \vartheta_{,k} \partial \vartheta_{,i}} - \Lambda \frac{\partial \varepsilon}{\partial \vartheta_{,k} \partial \vartheta_{,i}}\right) = \frac{\partial \Lambda}{\partial \vartheta} \frac{\partial q_i}{\partial \vartheta_{,k}}.$$

Therefore we obviously have

$$\frac{\partial q_{[i}}{\partial \vartheta_{,j]}} = 0 \quad \text{or} \quad K_{[ij]} = 0,$$

i.e. <u>the tensor of heat conductivity is symmetric.</u>

We may notice that this result is critically dependent on the fact that I have included $\dot{\vartheta}$ among the variables in the constitutive relations and that is also true for the prediction of finite speeds for temperature waves. Since the symmetry of the tensor of heat conductivity is widely believed to be an experimental fact and since a finite speed of temperature waves is certainly desirable on the grounds of general physical principles, I conclude that $\dot{\vartheta}$ <u>must</u> be considered as a variable

Part III

POROUS MATERIALS

1. Basic Concepts.

If the ideas of continuum mechanics are to be applied to porous solids whose pores are filled with fluids, it is nesessary to take a rather coarse view of such bodies. Essentially we must be able to say that every volume element in the body is simultaneously occupied by particles of the solid and the fluid and that implies that these volume elements must be big enough to embrace many pores.

Once we take that view, the continuum theory of mixtures offers itself for the description of this 'mixture' of a porous solid and a fluid because just like in the mechanical theory of mixtures, the

$$\text{densities } \rho_\alpha \text{ and the}$$
$$\text{motions } \chi_k^\alpha$$

of the constituents are fields to be determined here. The index α runs from 1 to 2 and I shall reserve $\alpha = 1$ for the fluid and $\alpha = 2$ for the solid. In this description all reference to pores would be missing and this is an unacceptable oversimplification which, if followed up, would lead to nonsensical results as I shall point out later.

While in the mechanical theory of mixtures the fields of ρ_α and χ_k^α are sufficient to describe the behaviour of a mixture, they are not in porous materials. Indeed, one way of obtaining a self consistent and physically reasonable theory of such materials is to define a field of porosity π and to regard this as a characteristic field in addition to the densities and the motions. The porosity is defined as the ratio of the total volume of the pores in a volume element and the volume of that element.

With this, the objective of a mechanical theory of porous materials may be defined as the determination of the fields of

$$\begin{array}{ll} \text{densities} & \rho_\alpha^{(+)} \\ \text{motions} & \chi_k^\alpha \quad \text{and} \\ \text{porosity} & \pi \end{array}$$

in a body.

(+) ρ_α here is the quotient of the mass of constituent α in a volume element and the volume of the element; it is not the true density of the fluid within the pores or the true density of the solid which will be introduced later and denoted by σ_α so that $\rho_1 = \sigma_1 \pi$ and $\rho_\alpha = \sigma_\alpha (1-\pi)$.

2. The Field Equations.

Having stated the problem it now remains to find equations by which the fields ρ_α, χ_k^α and π can be determined. To solve (say) an initial value problem in these 9 fields we need 9 differential equations as field equations and following the lead of continuum mechanics and thermodynamics we base these on the equations of balance of partial masses, partial momenta and energy. These equations of balance have the following form for a volume V with the boundary F that contains a moving singular surface φ :

$$\int_V \frac{\partial \rho_\alpha}{\partial t} dV + \oiint_F \rho_\alpha \vartheta_i^\alpha dF_i - \iint_\varphi [\rho_\alpha] u_i dF = 0 \qquad (\alpha = 1, 2)$$

$$\int_V \frac{\partial \rho_\alpha \vartheta_j^\alpha}{\partial t} dV + \oiint_F (\rho_\alpha \vartheta_j^\alpha \vartheta_i^\alpha - t_{ji}^\alpha) dF_i - \iint_\varphi [\rho_\alpha \vartheta_j^\alpha] u_i dF =$$

$$= \int (\rho_\alpha b_j^\alpha + m_j^\alpha) dV \qquad (\alpha = 1, 2)$$

$$\int_V \frac{\partial \rho(\varepsilon + \frac{1}{2}\vartheta^2)}{\partial t} dV + \oiint_F \left\{ \rho\left(\varepsilon + \frac{1}{2}\vartheta^2\right)\vartheta_i - t_{ji}\vartheta_j + q_i \right\} dF_i +$$

$$- \iint_\varphi \left[\rho\left(\varepsilon + \frac{1}{2}\vartheta^2\right)\right] u_i dF_i = \int \rho r \, dV$$

ρ is defined as $\rho_1 + \rho_2$. $\vartheta_i^\alpha \equiv \frac{\partial \chi_i^\alpha}{\partial t}$ is the velocity of constituent α, $\vartheta_i \equiv \sum_{\alpha=1}^{2} \frac{\rho_\alpha}{\rho} \vartheta_i^\alpha$ the barycentric velocity and u_i is the normal velocity of the singular surface; b_j^α and r

are the specific external body forces on constituent α and the specific radiation supply respectively. t_{ij}^{α} denotes the partial stresses and m_j^{α} the interaction force density between the constituents; t_{ij} is defined in terms of t_{ij}^{α} by the relation

$$t_{ij} \equiv \sum_{\alpha=1}^{2} \left(t_{ij}^{\alpha} - \rho_{\alpha} (\vartheta_i^{\alpha} - \vartheta_i)(\vartheta_j^{\alpha} - \vartheta_j) \right).$$

ε is the specific internal energy and q_i its flux. $[\psi]$ denotes the jump of ψ across the singular surface.

The law of conservation of total momentum is expressed by the requirement that the production of total momentum $\sum_{\alpha=1}^{2} m_j^{\alpha}$ be zero.

The above equations of balance are equivalent to the local equations

$$\frac{\partial \rho_{\alpha}}{\partial t} + (\rho_{\alpha} \vartheta_i^{\alpha})_{,i} = 0 \qquad (\alpha = 1,2)$$

(I) $\qquad \dfrac{\partial \rho_{\alpha} \vartheta_j^{\alpha}}{\partial t} + (\rho_{\alpha} \vartheta_j^{\alpha} \vartheta_i^{\alpha} - t_{ji}^{\alpha})_{,i} = \rho^{\alpha} b_j^{\alpha} + m_j^{\alpha} \quad \sum_{\alpha=1}^{2} m_j^{\alpha} = 0 \quad (\alpha = 1,2)$

$$\frac{\partial \rho \varepsilon}{\partial t} + (\rho \varepsilon \vartheta_i + q_i)_{,i} = t_{ij} \frac{\partial \vartheta_i}{\partial x_i} + \rho r$$

for all points that do not lie on the singular surface, while for points on the singular surface they are equivalent to the jump conditions

$$[\rho_\alpha(\vartheta_i^\alpha - u_i)e_i] = 0, \qquad (\alpha = 1,2)$$

$$[\rho_\alpha \vartheta_j^\alpha(\vartheta_i^\alpha - u_i)e_i] - [t_{ji}^\alpha e_i] = 0, \qquad (\alpha = 1,2)$$

$$[\rho(\varepsilon + \tfrac{1}{2}\vartheta^2)(\vartheta_i - u_i)e_i] - [t_{ji}\vartheta_j e_i] + [q_i e_i] = 0$$

under rather general conditions. e_i is the normal vector to φ that points in the direction of u_i.

Althought we now have the required number of equations, these cannot in the present form be considered as field equations for ρ_α, χ_k^α and π. What is needed are constitutive equations that relate t_{ij}^α, m_i^α, ε and q_i to the field of densities, motions and porosity in a materially dependent manner. In general t_{ij}^α, m_j^α, ε and q_i at one point x_n and at time t could depend on the histories of ρ_α, χ_k^α and π in the whole body. Here, however, I consider materials whose properties $t_{ij}^\alpha(x_n, t)$, $m_j^\alpha(x_n, t)$, $\varepsilon(x_n, t)$ and $q_i(x_n, t)$ depend only on the densities ρ_α and the porosity π at the place x_n and time and on the velocities $\vartheta_k^\beta(x_n, t)$ and the deformation gradients $F_{kA}^\beta(x_n, t) \equiv \dfrac{\partial \chi_k^\beta(X_B^\beta, t)}{\partial X_A^\beta}$ with respect to the reference configuration K_β in which the particles of constituent β occupied the positions X_B^β. If in addition the solid is assumed to be isotropic, one can show (+) from the principle of objectivity and the isotropy requirements that the constitutive relations have the general form

(II)
$$t^\alpha_{ij} = \hat{t}^\alpha_{ij}(V_k, \varrho_1, B^2_{ij}, \pi), \qquad (\alpha = 1,2)$$

$$m^1_j = \hat{m}^1_j(V_k, \varrho_1, B^2_{ij}, \pi), \qquad m^2_j = -m^1_j$$

$$\varepsilon = \hat{\varepsilon}(V_k, \varrho_1, B^2_{ij}, \pi),$$

$$q_i = \hat{q}_i(V_k, \varrho_1, B^2_{ij}, \pi), \quad ^{(++)}$$

where $B^2_{ij} = F^2_{iA} F^2_{jA}$ is the left Cauchy Green tensor and $V_k \equiv \vartheta^1_k - \vartheta^2_k$ is the relative velocity of the constituents. The constitutive functions $\hat{t}^\alpha_{ij}$, $\hat{m}^1_j$, $\hat{q}_i$ and $\hat{\varepsilon}$ are isotropic tensorial, vectorial and scalar functions of their variables with respect to the Euclidean group of transformations.

If the constitutive relations (II) are introduced in the local equations of balance (I) what results is a determinate system of field equations from which we may hope to predict the

---
(+) I. MÜLLER, A Macroscopic Model for the Flow of a Fluid through a Porous Solid, submitted to ZAMP.
(++) For generality I have employed the principle of equipresence here, according to which all constitutive quantities may depend on the variables.

fields ρ_α, χ_k^α and π as solutions of a well posed boundary and initial value problem.

3. Some Special Cases.

It is easily appreciated from the above arguments that the field equations just derived are far from general and yet they are much too general to be of direct use to people interested in soil mechanics or hydrology. These people usually cut the problem to size by assuming that

$$t_{ij}^1 = -p^1(\rho_1, \pi)\delta_{ij}$$

$$m_j^1 = -RV_j ,$$

where p^1 is called the fluid pressure and R the resistivity constant. I shall henceforth also use these simplifying assumptions. Further important simplifications of the problem are made possible by the observation that most of the materials of interest are not capable of arbitrary motions or density changes or porosity changes; rather these fields are constraint and I shall now proceed to discuss such constraints and indicate the resulting simplifications.

a.) Incompressible Fluid

Remembering that $\rho_1 = \sigma_1 \pi$ where σ_1 is the density of the fluid within the pores, we see that incompressibili̲ty of the fluid means

$$G_1 = \text{const}.$$

so that ρ_1 and π are proportional. This constraint leaves us therefore with one field less to consider and we may feel free to disregard the balance of internal energy as a field equation. Therefore we have the set of equations

(III)
$$\frac{\partial \rho_\alpha}{\partial t} + (\rho_\alpha \vartheta_i^\alpha)_{,i} = 0$$

$$\frac{\partial \rho_\alpha \vartheta_j^\alpha}{\partial t} + (\rho_\alpha \vartheta_j^\alpha \vartheta_i^\alpha - t_{ji}^\alpha)_{,i} = \rho_\alpha b_j^\alpha + m_j^\alpha$$

$(\alpha = 1, 2)$

which must be supplemented by the constitutive relations

(IV)
$$t_{ij}^1 = -p^1(\rho_1)\delta_{ij} \quad,$$

$$t_{ij}^2 = \hat{t}_{ij}^2(V_k, \rho_1, B_{ij}^2) \quad,$$

$$m_j^1 = -RV_j \quad, \qquad m_j^2 = -m_j^1$$

to give a set of 8 equations for the 8 fields ρ_α, χ_k^α.

b.) Constant Porosity

If the porosity is constant, we have a formally

very similar situation as in section a.). In fact the equations (III) and (IV) describe this case as well, if only it is understood that the symbols p_1 and $\hat{t}^2_{ij}$ denote different functions now.

c.) Incompressible Fluid and Constant Porosity

We now consider the two constraints

$$\sigma_1 = \text{const.} \quad \text{and} \quad \pi = \text{const.}$$

so that $\rho_1 = \text{const.}$. Here the eight field equations (III) with (IV) would be an overdeterminate set of field equations for the seven fields $\rho_2, \chi^1_k, \chi^2_k$. I shall follow the customary procedure in this case by assuming that the trace of t^1_{ij} is not determined by a constitutive equation, i.e. that the fluid pressure, which in this case I denote by P is left arbitrary. Thus among (III) we have seven equations that determine $\rho_2, \chi^1_k, \chi^2_k$ and the remaining one may be considered as an equation that determines P, if boundary values of P are prescribed.

d.) Rigid Solid and Constant Porosity

If the solid is rigid and has constant porosity, we have $\pi = \text{const.}$ and the motion χ^2_k can only be a rigid rotation and translation so that $B^2_{ij} = 1$. Therefore no field equations are needed for π and χ^2_i and we may delete the balance of internal energy and the balance of momentum of the solid from the set (I). The balance of mass

$$\frac{\partial \rho_2}{\partial t} + (\rho_2 v^2_i)_{,i} = 0$$

in this case merely expresses that the density of the solid particles is constant and may therefore also be deleted. This leaves us with two equations:

(V)
$$\frac{\partial \rho_1}{\partial t} + (\rho_1 \vartheta_i^1)_{,i} = 0$$

$$\frac{\partial \rho_1 \vartheta_j^1}{\partial t} + (\rho_1 \vartheta_j^1 \vartheta_i^1 - t_{ij}^1)_{,i} = \rho_1 b_j^1 + m_j^1$$

which, with the constitutive relations

(VI)
$$t_{ij}^1 = -p'(\rho_1) \delta_{ij}$$

$$m_j^1 = -R V_j$$

furnish a set of equations for the determination of ρ_1 and χ_k^1.

e.) <u>Incompressible Fluid in a Rigid Solid with Constant Porosity</u>

Like in c.), when the fluid is incompressible and the porosity constant, so that $\rho_1 = $ const., we do not have to specify the trace of t_{ij}^1 by a constitutive relation. From (V) and (VI) we infer that the field equations that govern this case, have the form

(VII)
$$\vartheta_{i,i}^1 = 0,$$

$$\rho_1 \frac{\partial \vartheta_j^1}{\partial t} + \rho \vartheta_i^1 \vartheta_{j,i}^1 - t_{ji,i}^1 = \rho_1 b_j^1 + m_j^1$$

with

$$t'_{ji} = -P \delta_{ij}$$
$$m'_j = -R V_j$$
(VIII)

From these equations we may determine the fields of χ'_k and P.

4. Darcy's Law.

Much of the literature on porous solids is concerned with Darcy's law which gives an empirical linear relation between the relative velocity V_j and the fluid pressure p_1. Clearly, if we had such a relation in addition to our set of equations (I), we should not in general expect a solution. What is then the role of Dancy's law in the theory presented here? The answer to this question is simple enough if we know how Fick's law of diffusion in homogeneous mixtures and Ohm's law of diffusion of electrically charged constituents are interpreted in systematic theories of mixtures: All these "laws" are simplified versions of the equations of balance of momenta.

In the general case of an unconstrained material where the field equations are given by (I) and (II) on p. 38 and 40 we can hardly derive Darcy's law from the equations of balance of momenta without the most sweeping assumptions. But in the case of a rigid solid with constant porosity the balance of momentum of the fluid has the form (see (V) and (VI) on p.44)

$$\rho_1 \dot{\vartheta}'_j + \underline{p_{1,j}} = \rho_1 b'_j - \underline{RV'_j} \; ,$$

where

$$\dot{\vartheta}'_j \equiv \frac{\partial \vartheta'_j}{\partial t} + \vartheta'_i \vartheta'_{j,i}$$

is the acceleration of the fluid.

The underlined part of the equation is Darcy's law and we conclude therefore that Darcy's law holds in an acceleration-free flow of a fluid through a rigid solid with constant porosity when the body force is ignored and when the simplifying constitutive assumptions on t'_{ij} and m'_j of p. 41 are accepted.

5. Motivation for Taking the Porosity as a Variable.

Obviously the problem of showing that the porosity ought to be taken into account as a variable would be solved, if we could show that this is the case for but one special case. Let us therefore consider a time independent process in a fluid and a solid with non-uniform porosity which are both at rest. Without essential loss of generality we may ignore body forces and thus obtain from (I) and (II) with $t'_{ij} = - p_1(\rho_1, \pi) \delta_{ij}$ and $m'_j = - RV'_j$ as on p. 41

$$p_{1,j} = 0 \; .$$

Interface boundary conditions

From this relation and the intuitive expectation that the fluid pressure within the pores be uniform for the process considered we conclude that p_1 must be interpreted as the fluid pressure within the pores.

Also on intuitive grounds we expect σ_1 to be uniform in this process and therefore $\varrho_1 = \sigma_1 \pi$ is certainly not uniform since π is not. Now suppose p_1 would not depend on π but only on ϱ_1; is that case we should be unable to satisfy the relation $p_{1,j} = 0$ in general. One way to put that right is to consider π as a variable, at least in p_1.

Actually in the above special case we should expect that p_1 is a function of $\frac{\varrho_1}{\pi}$. In general this may or may not be so; in any case π must enter the arguments of our constitutive relations along with ϱ_α and we are on the safe side if we assume that ϱ_α and π enter as independent variables.

6. Interface Boundary Conditions.

Hydrologists and soil engineers are quite often interested in what happens when a fluid traverses the boundary between two different porous materials I and II, and since there is some confusion as to the right boundary conditions, I proceed to derive these.

So as not to complicate the argument with boundary conditions on little suggestive quantities like the flux of internal energy, I choose to consider the special case that both solids have a constant but different porosity. As I pointed out in chapter 3. b.) the field equations for this case are (III) and (IV) on p. 42 and correspondingly the boundary conditions are

$$[\rho_\alpha (\vartheta_i^\alpha - u_i) e_i] = 0 ,$$

$$(\alpha = 1,2)$$

$$[\rho_\alpha \vartheta_j^\alpha (\vartheta_i^\alpha - u_i) e_i] - [t_{ji}^\alpha e_i] = 0 .$$

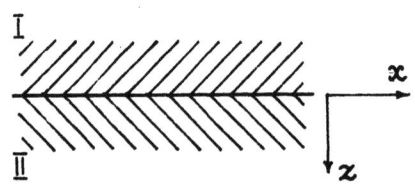

Clearly the singular surface is a material surface of the two solids so that

$$u_i e_i = \vartheta_i^2 e_i$$

and with this the above jump conditions reduce to

$$[\rho_1 (\vartheta_i^1 - \vartheta_i^2) e_i] = 0 ,$$

$$[\vartheta_j^1] \rho_1 (\vartheta_i^1 - \vartheta_i^2) e_i + [p_1] e_j = 0 ,$$

$$[t_{ji}^2 e_i] = 0 ,$$

where t_{ij}^1 has been assumed to be given by $-p_1 \delta_{ij}$ as on p. 41 Jump conditions for the fluid velocity follow easily from the second one of these relations:

Interface boundary conditions

$$[\vartheta_\|^1] = 0 \quad , \quad [\vartheta_\perp^1] = -\frac{[p_1]}{Q} \quad ^{(+)}$$

where $\vartheta_\|^1$ and $\vartheta_\perp^1$ are the tangential and normal components of the fluid velocity at the interface and $Q \equiv \varrho_1(\vartheta_j^1 - \vartheta_j^2)e_j$ is the mass flux of fluid through the interface.

We conclude that <u>while the tangential component of the fluid velocity is continuous, the normal component suffers a jump whose size is determined by the mass flux through the surface and by the jump in pressure.</u>

This conclusion, trivial as it is, would surprise almost every hydrologist, since in hydrology everybody calculates fairly successfully with boundary conditions that read: The normal component of ϑ_i^1 is continuous, while the ratio of the tangential components $\vartheta_\|^1$ in the two media is equal to the reciprocal of the ratio of the resistivities.

In a recent paper P.A.C. Raats (++) has given an argument that indicates why these faulty boundary conditions have worked fairly well for people in hydrology. Raats considers an incompressible fluid in rigid solids with the uniform porosity so that according to (VII) and (VIII) the field equations have the form

---

(+) If the fluid is incompressible and the porosity constant, I write P instead of p_1 as a reminder that the pressure is then not a constitutive quantity.
(++) P.A.C. Raats. Submitted to Arch. Rat.Mech. Anal.

$$\vartheta^1_{i,i} = 0$$

$$\rho_1 \frac{\partial \vartheta^1_j}{\partial t} + \rho_1 \vartheta^1_i \vartheta^1_{j,i} + P_{,j} = -R(\vartheta^1_j - \vartheta^2_j) \, .^{(+)}$$

in both solids. In such a material Raats considers a special two dimensional flow in the x, z plane that is characterized by the fields

ϑ^1_j time independent and uniform in I

ϑ^1_j time independent and independent of x in II

ϑ^2_j identically zero in I and II

The solution of this problem in II is easily seen to be

$$\vartheta^1_\parallel = -\frac{(P_{,x})^{II}}{R^{II}} + c_1 e^{-\frac{R^{II}}{Q}z}$$

$$\vartheta^1_\perp = c_2 \, ,$$

where c_1 and c_2 are constants of integration that can be expressed in terms of $(\vartheta^1_\parallel)^I$, $(\vartheta^1_\perp)^I$, $[P]$ and Q by virtue of the jump conditions $[\vartheta^1_\parallel] = 0$ and $[\vartheta^1_\perp] = -\frac{[P]}{Q}$:

$$c_1 = (\vartheta^1_\parallel)^I + \frac{(P_{,x})^{II}}{R^{II}} \, , \quad c_2 = (\vartheta^1_\perp)^I + \frac{[P]}{Q} \, .$$

(+) While Raats allows for conservative body forces, I do not in this brief note.

In the above solution, use has been made of the fact that $P_{,x}$ is uniform in II and since $[P]$ is independent of x, we have $P_{,x} = (P_{,x})^{II} = (P_{,x})^{I}$ which according to the momentum balance in I is equal to $-R^{I}(\vartheta_{//}^{1})^{I}$.

Thus we obtain in II

$$\vartheta_{//}^{1} = \frac{R^{I}}{R^{II}}(\vartheta_{//}^{1})^{I} + (\vartheta_{//}^{1})^{I}\left(1 - \frac{R^{I}}{R^{II}}\right)e^{-\frac{R^{II}}{Q}z},$$

$$\vartheta_{\perp}^{1} = (\vartheta_{\perp}^{1})^{I} + \frac{[P]}{Q}.$$

We thus see that the value $\frac{R^{I}}{R^{II}}(\vartheta_{//}^{1})^{I}$ which hydrologists assume is equal to $(\vartheta_{//}^{1})^{II}$, is in fact the limiting value of $\vartheta_{//}^{1}$ in II for great z. Dr. Raats justifies the usage of the rule $(\vartheta_{//}^{1})^{II} = \frac{R^{I}}{R^{II}}(\vartheta_{//}^{1})^{I}$ by showing that in most instances hydrologists are concerned with, $\frac{R^{II}}{Q}$ is sufficiently big that the exponential term above can be neglected in any appreciable distance from the interface boundary.

The second boundary condition commonly used in soil mechanics is also not strictly correct. In fact $[\vartheta_{\perp}^{1}] = 0$ would only hold if $[P] = 0$, but from our calculation we see that

$$[P] = \left(1 - \frac{\pi^{I}}{\pi^{II}}\right)Q^{2}.$$

Contents

	Page
Preface..	3
Part I. THE COLDNESS, A UNIVERSAL FUNCTION IN THERMO-DYNAMICS OF SIMPLE HEAT CONDUCTING FLUIDS.....	5
1. Thermodynamic Processes............................	5
2. Restrictive Principles.............................	7
3. Restrictions from the Entropy Inequality on the Constitutive Functions.............................	8
4. The Coldness......................................	15
5. The Speed of Heat Propagation.....................	18
Part II. ON HEAT CONDUCTION IN A RIGID BODY...........	21
1. General Ideas.....................................	21
2. Restrictions from the Entropy Inequality on the Constitutive Relations for Rigid Solids............	23
3. Isotropic Rigid Solids and the Coldness...........	26
4. Temperature Waves in Isotropic Rigid Solids.......	29
5. Symmetry of the Tensor of Heat Conductivity in Anisotropic Rigid Solids...........................	31
Part III. POROUS MATERIALS............................	35
1. Basic Concepts....................................	35
2. The Field Equations...............................	37
3. Some Special Cases................................	41
4. Darcy's Law......................................	45
5. Motivation for taking the Porosity as a Variable...	46
6. Interface Boundary Conditions.....................	47

MIX
Papier aus verantwortungsvollen Quellen
Paper from responsible sources
FSC® C105338

If you have any concerns about our products,
you can contact us on
ProductSafety@springernature.com

In case Publisher is established outside the EU,
the EU authorized representative is:
Springer Nature Customer Service Center GmbH
Europaplatz 3, 69115 Heidelberg, Germany

Printed by Libri Plureos GmbH
in Hamburg, Germany